LANDSCAPE RESILIENCE FRAMEWORK

Operationalizing ecological resilience at the landscape scale

San Francisco Estuary Institute
Erin Beller
April Robinson
Robin Grossinger
Letitia Grenier

June 2015

SFEI
A S C

SFEI Publication # 752

Introduction

As human populations expand and our demands on the environment increase, we must wrestle with the question of how we can create and sustain diverse, healthy ecosystems across the landscapes we inhabit. How can we design landscapes that provide meaningful and lasting benefits to both people and wildlife? How can we sustain biodiverse, healthy ecosystems, from our cities to our wildlands, with the capacity to persist and evolve over time?

These questions, always challenging, have become even more so in the face of the rapid environmental changes that are anticipated over the coming century, particularly stressors associated with climate change and development. As we plan for impacts that are likely to be unpredictable or unprecedented, it is increasingly crucial that we support ecosystems flexible enough to adjust and reassemble, maintaining biodiversity and ecological functions in response to significant changes.

In this context, the concept of resilience, with its explicit focus on creating systems that are robust enough to persist and adapt over the long run, has emerged as a particularly relevant way to manage ecosystems for an uncertain future. There are tremendous opportunities for us – as natural resource managers, conservation scientists, urban planners, landscape architects, and many others – to strengthen the resilience of our landscapes and help promote ecosystems, habitats, and species that are likely to successfully adapt and thrive over time. While resilience-based management has widespread appeal and potential, however, it is notoriously difficult to operationalize.

The Landscape Resilience Framework presented here is designed to facilitate application of resilience principles to ecosystem management by detailing the seven dimensions of a landscape that contribute to resilience. It represents a synthesis of thinking across empirical ecological studies and social-ecological resilience theory, and was reviewed by a team of expert advisors. Our goal was to create a concise and comprehensive set of key considerations that could be integrated into identifying on-the-ground actions across urban design, conservation planning, and ecosystem management that would contribute to resilient future landscapes.

Background

WHAT IS RESILIENCE?

Resilience as applied to ecosystems was first defined as a measure of a system's ability to absorb change and persist after a perturbation (Holling 1973). Over the past decades, however, application of the term has expanded to encompass social, economic, and infrastructure systems in addition to ecological systems (Brand and Jax 2007, Curtin and Parker 2014; see right). Today, the concept of resilience is perhaps most frequently applied to social-ecological systems: that is, interrelated networks of ecosystems, institutions, actors, and species (cf. Berkes et al. 2003, Adger et al. 2005).

The focus of this framework is ecological resilience at a landscape scale, or "landscape resilience," as one dimension of resilience within social-ecological systems. For our purposes, we define landscape resilience as the ability of a landscape to sustain desired ecological functions, robust native biodiversity, and critical landscape processes over time, under changing conditions, and despite multiple stressors and uncertainties. While social and ecological systems are inextricably linked and all aspects of social-ecological resilience are important to understand, developing a robust understanding of the mechanisms of ecological resilience in and of themselves is an essential step in applying the broader concept (Standish et al. 2014).

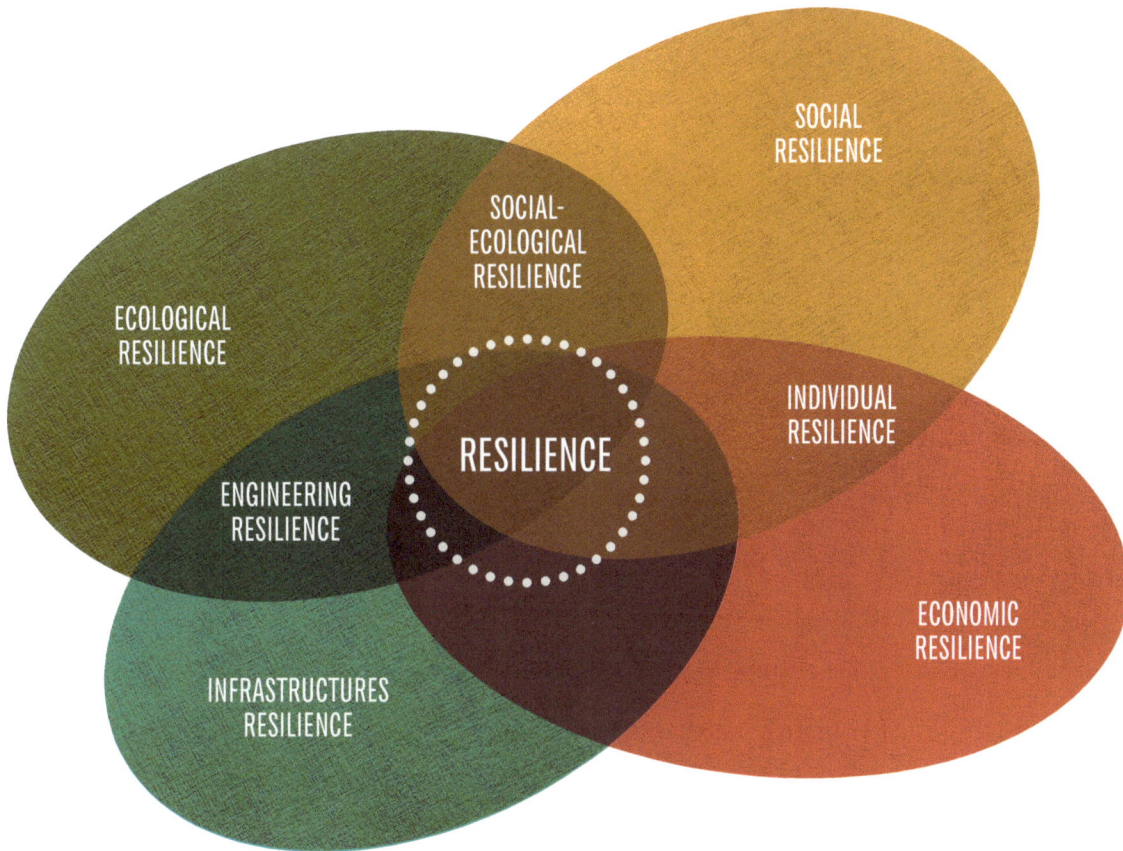

While the concept of resilience can be applied to a broad array of systems, the Landscape Resilience Framework presented here focuses on ecological resilience. (adapted from Chelleri and Olazabal 2011)

RESILIENCE "OF WHAT, TO WHAT"?

In order to manage for landscape resilience, one must first establish which system attributes are desired to be resilient to which environmental stressors or drivers of change - that is, one must determine the resilience "of what, to what?" (Carpenter et al. 2001, Zavaleta and Chapin 2010). At the broadest scale, the goal of the framework is to identify landscape elements that sustain biodiverse, ecologically functional landscapes in the context of climate change and other anthropogenic stressors over the coming century and beyond (see boxed text, page 8).

By "biodiverse, ecologically functional landscapes," we mean landscapes that support a diversity and abundance of life, along with the ecosystem components (species, habitats, and processes) necessary to sustain it in perpetuity. Plants and animals are of particular concern because they are often better studied and garner more management attention than other species; however, we include other taxa to the degree they support these species. Our emphasis is on primarily native species that are adapted to local ecological conditions, though non-native species are valued to the degree that they support native species or desired ecological functions, or when replacement by analogous native species is infeasible or undesirable. In addition to currently present species, it also includes extirpated species that might be recoverable as well as species whose ranges may shift to include a given area in the future.

By "climate change and other anthropogenic stressors," we are referring to an array of ongoing and projected future drivers of change that are likely to impact ecosystems. These include both chronic stressors (such as sea-level rise) and events (such as floods), as well as short-term and long-term stressors. In particular, we emphasize impacts associated with climate change such as increased temperatures, increased drying, and sea-level rise; modifications of physical and biological processes such as changes to freshwater flows or sediment supply; and modifications to landscapes, such as urbanization or habitat loss.

Since the specific stressors and aspects of biodiversity that are most important are highly context dependent, applying the resilience framework requires an extensive understanding of place. This includes an understanding not only of current conditions, but also of past conditions and future projections – how the landscape has changed, and is likely to change, over time. The realities of different settings offer distinct opportunities and constraints on the ecological functions that can be maintained.

Applying the resilience framework is also scale dependent. While at a regional landscape scale we focus generally on resilience of biodiversity to multiple stressors, at smaller spatial scales we emphasize particular ecological functions and processes nested within biodiversity that are most relevant to a specific location (e.g., movement corridors for large mammals, nesting habitat for neo-tropical migrant songbirds) and the particular stressors that pose the most severe or urgent threats (e.g., increased urban development, spread of invasive species).

RESILIENCE "OF WHAT"?

Our focus is on the resilience of biodiversity and the ecological functions needed to sustain it in perpetuity. Biodiversity includes a diversity and abundance of life at all levels, from genes to ecosystems. Ecological functions are the ways in which components of an ecosystem (species, habitats, biological processes) support life. More detail about what we consider under these terms is provided below.

SPECIES OF INTEREST:

- Primarily **native plants and animals**; other taxa to the degree that they support these species. Includes extirpated species that might be reintroduced and nearby species whose ranges may shift to include this area in the future

- **Non-native species** to the degree that they support native species or desired ecological functions, or when replacement by analogous native species is infeasible or undesirable

ASPECTS OF HABITATS:

- Water availability (maintained through a diversity of aquatic and wetland habitats, hydrologic processes, and the water cycle)

- Prey availability (e.g., acorns, emergent insects, phytoplankton)

- Habitats/conditions that permit foraging (e.g., raptor perches, mudflats for shorebirds, appropriate turbidity for fish to forage successfully)

- Sites for reproduction (e.g., nest sites, spawning beds)

- Cover/refuge to escape predators and physical stressors (such as flooding, fire, extreme temperatures)

- Conditions/features that ameliorate physical stressors (e.g., wave barriers, dense canopies to keep creek temperatures cool, mycorrhizal associations that increase plant tolerance)

- Space for daily movement

- Connection between areas supporting different aspects of species life history

- Environmental conditions that fall within the physiological tolerance of the species (e.g., temperature, salinity)

BIOLOGICAL PROCESSES:

- Modification of the physical environment (e.g., trampling, burrowing, shading, altering flows)
- Population dynamics (e.g., reproductive success, recruitment)
- Movement (e.g., dispersal, gene flow, migration)
- Adaptive evolution and co-evolution (e.g., pollination, predator-prey)
- Food web dynamics (e.g., primary productivity, predation, decomposition, nutrient cycling)
- Competition (e.g., space, reproduction, food)
- Symbiosis (e.g., mutualism, commensalism, parasitism)
- Restructuring of biological communities (e.g., invasion/introduction, niche differentiation, extinction)

RESILIENCE "TO WHAT"?

Ongoing and projected future stressors that are likely to impact ecosystems, including:

- **Climate change**, including increased frequency of extreme events such as floods and fires, increased temperature, increased drying, changes in precipitation patterns, and sea-level rise
- **Modification of physical and biological processes**, including changes to freshwater and tidal flows; changes in sediment supply, demand, and transport; and invasion by non-native species
- **Landscape modification**, including urbanization and increased development, habitat loss and degradation, human-subsidized predators, contaminants and pollution, urban heat island effects, and human disturbance

WHY LANDSCAPE RESILIENCE?

Landscapes span a range of land covers and habitat types, from rural and wilderness areas to cities and suburban communities. No matter the setting, each landscape includes multiple habitats or ecosystems interacting across large spatial scales. Our focus on ecological resilience at a landscape scale stresses the importance of considering resilience over the scales at which the human actions and biological and physical processes needed to sustain biodiversity and ecological functions occur.

By emphasizing resilience, we aim to envision landscapes that support high levels of desired ecological functions and biodiversity over time, even as some state changes and transformations in the structure, form and condition of the landscape occur. This is in contrast to existing conditions in many systems, which are often degraded and have low capacity for persistence and adaptation. Applying the resilience framework will therefore require not only protecting and conserving important aspects of existing landscapes that contribute to resilience, but also active management and restoration activities that address past and current stressors and that look to the future to consider which additional features will bolster landscape resilience.

WHAT ABOUT ECOSYSTEM SERVICES?

While this framework focuses exclusively on the principles that contribute to the resilience of biodiversity and ecological functions, this is only one dimension of how resilience is applied to cultural and ecological systems. Other factors not considered in this framework, notably the resilience of ecosystem services and human infrastructures and institutions, are equally important considerations when assessing and improving resilience in social-ecological systems.

Our emphasis on biodiversity and ecological function endeavors to fill a gap in the resilience literature. While existing documents explore the elements that promote the resilience of ecosystem services (cf. Resilience Alliance 2010; Biggs et al. 2012, 2015), to our knowledge no comparable framework yet exists for ecological functions. However, there are clearly strong connections between ecological functions (that benefit non-human species, such as water or prey availability, cover, or nest sites) and associated ecosystem services (that benefit people and society, such as a clean water supply, groundwater storage, or recreation), as well as substantial overlaps in the resilience principles contributing to each. Thus while we are not explicitly focusing here on benefits to people and society, managing for the resilience of ecological functions will inevitably yield numerous co-benefits, in addition to some trade-offs. We see this document as complementary to existing efforts that have a heightened societal focus; ultimately, we envision this document could be incorporated into broader social-ecological resilience frameworks to add detail and depth to the ecological dimension of social-ecological resilience.

East Palo Alto, California (imagery courtesy of Google Earth)

About this document

In the following pages, we highlight seven principles of landscape resilience, along with the elements within each principle most relevant to planning, restoration, conservation, and management. We discuss how each principle contributes to landscape resilience and provide examples of how they might be integrated into a vision for a resilient landscape.

This Landscape Resilience Framework was developed as part of Resilient Silicon Valley, a project funded by a charitable contribution from Google's Ecology Program, to guide the creation of science-based visions for ecosystem health and resilience in Silicon Valley and beyond. We are currently in the process of applying these principles to Silicon Valley to produce a vision for landscape resilience (see resilientsv.sfei.org for more information). This project will provide an example of how these principles can be applied locally to develop a vision for landscape resilience for a particular location.

This document reflects the best available science derived and synthesized from case-studies and theory in the resilience literature, as well as input from an international Resilience Advisory Team (see acknowledgments). It is designed to provide the structure needed to comprehensively incorporate the key concepts for building landscape resilience into restoration and management planning, as is currently being piloted for the Silicon Valley, California. However, there is inevitably uncertainty in the mechanisms of resilience: resilience science is continually evolving, and there are limits on what has been empirically studied. We view this as a living document and expect it to be refined over time as more empirical studies and pilot projects are completed and resilience theory evolves. In the meantime, we hope it can serve as a catalyst for discussions about how we want to shape our landscapes in the coming century to prioritize ecological resilience.

Principles of Landscape Resilience

We have identified seven principles of landscape resilience: setting, process, connectivity, diversity/complexity, redundancy, scale, and people. Together, these seven principles encapsulate the most important considerations when planning for landscape resilience.

Each of the seven principles contributes to resilience in slightly different ways. Setting and process, cover the physical, ecological, and social/cultural context of a particular landscape, and therefore drive the appropriate expression of the other principles for that place. Connectivity, diversity/complexity, and redundancy capture the distribution, configuration, and abundance of species and habitats across a landscape. Scale and people mediate each of the other principles by influencing how actions can be implemented in a landscape.

There are multiple overlaps, synergies, and trade-offs within and between the principles. For example, stream flows that connect channels to floodplains encompass elements of both connectivity and process; pools of varying depths exemplify both diversity and redundancy. With limited space and resources, there are often trade-offs between principles: for example, between diversity/complexity and redundancy (e.g., choosing between multiples of the same habitat patch versus different types of habitat) or redundancy and connectivity (e.g., linking habitat patches to promote species movement versus keeping them isolated to promote diversity). Implicit in the framework is that for many of the principles, more is not necessarily better: for connectivity in particular, low levels of connectivity are most appropriate in some situations (e.g., low hydrologic connectivity for intermittent streams, or low genetic connectivity for isolated populations).

1	**SETTING** ········➤	Unique geophysical, biological, and cultural aspects of a landscape that determine potential constraints and opportunities for resilience
2	**PROCESS** ········➤	Physical, biological, and chemical drivers, events, and processes that create and sustain landscapes over time
3	**CONNECTIVITY** ········➤	Linkages between habitats, processes, and populations that enable movement of materials and organisms
4	**DIVERSITY & COMPLEXITY** ········➤	Richness in the variety, distribution, and spatial configuration of landscape features that provide a range of options for species
5	**REDUNDANCY** ········➤	Multiple similar or overlapping elements or functions within a landscape that promote diversity and provide insurance against loss
6	**SCALE** ········➤	The spatial extent and time frame at which landscapes operate that allows species, processes, and functions to persist
7	**PEOPLE** ········➤	The individuals, communities, and institutions that shape and steward landscapes

SETTING determines the constraints and opportunities within a landscape

What is it? Setting encapsulates the unique geophysical, biological, and cultural aspects of a particular landscape. At a fundamental level, setting dictates which ecological functions, processes, and biological communities are inherently possible and appropriate for a given landscape to sustain. Key to setting is an understanding of not just current conditions, but also system trajectories: which facets of setting have persisted over time, and which have changed? How are these elements likely to shift in the future?

What are the key elements?

GEOPHYSICAL CONTEXT: Underlying geology, soils, and topography

ECOLOGICAL CONTEXT: Ecological assemblies; dominant or unique vegetative communities that distinctively characterize the landscape, including those that characterized the landscape in the past; coevolved relationships

HISTORICAL AND CULTURAL CONTEXT: How the landscape has changed over time – which elements have persisted or disappeared, and why; opportunities provided by anticipated future land use trajectories

CRITICAL RESOURCES: Resources required for the persistence of desired ecological functions but currently limited within the landscape

How does it contribute to resilience?

Setting shapes how the other landscape resilience principles are applied in a given landscape. It encompasses the conditions under which native species evolved and to which they are adapted (ECOLOGICAL CONTEXT) and reveals enduring aspects of the physical context likely to persist even in the face of stressors (GEOPHYSICAL CONTEXT). Understanding local history and change over time can help us understand how the landscape responded to environmental stressors in the recent past, can illuminate landscape elements that have persisted over time that might contribute to resilience (e.g., remnant seed banks or intact wetland topography), and can inform potential future trajectories as infrastructure and landscapes are redesigned (HISTORICAL/CULTURAL CONTEXT). Setting determines which resources are limiting (e.g., availability of summer freshwater in Mediterranean climates); these resources may need to be explicitly incorporated into management and adaptation strategies in order to maintain desired ecological functions (CRITICAL RESOURCES).

EXAMPLES
from Resilient Silicon Valley vision:

- The landscape supports native vegetative communities (e.g., oak savanna and woodland, serpentine grassland, sycamore-alluvial woodland) (ECOLOGICAL CONTEXT)

- Groundwater levels are sufficient to maintain persistent, stratified summer stream pools for aquatic organisms and naturally perennial stream reaches (GEOPHYSICAL CONTEXT)

- Changes in salt production industry provide opportunities for baylands habitat restoration (HISTORICAL/CULTURAL CONTEXT)

- Nest boxes provide habitat for cavity-nesting birds in areas where nesting sites are limited due to the removal of older trees in oak woodlands (CRITICAL RESOURCES)

PROCESSES create and sustain landscapes in a dynamic way

What is it? Physical, biological, and chemical drivers, events, and processes shape landscapes at a variety of spatial and temporal scales. They contribute to the movement of materials in the landscape, help create and maintain habitats and habitat heterogeneity, and spatially organize ecological functions and communities.

What are the key elements?

SYSTEM DRIVERS: Large-scale forces such as climate and land use

DISTURBANCE REGIMES: Expected but unpredictable events, such as fires, floods, and droughts, that reset and create new habitats at certain frequencies and magnitudes

HABITAT-SUSTAINING PROCESSES: Dynamic physical processes, such as the transport of water and sediment, that sustain habitats

How does it contribute to resilience?

Physical drivers such as precipitation gradients interact with setting to determine which ecological functions are likely to be able to persist and whether species will be able to adapt to environmental change in a particular place (SYSTEM DRIVERS). Processes create complex, heterogeneous landscapes and habitats and support temporal variability in habitat. They also distribute resources such as water and sediment (DISTURBANCE REGIMES, HABITAT-SUSTAINING PROCESSES). It is worth noting that some events considered part of disturbance regimes, such as wildfire or floods, also have the potential to be stressors that must be managed for, depending on their magnitude, frequency, and context.

EXAMPLES
from Resilient Silicon Valley vision:

- Bayland habitats span key gradients in salinity (SYSTEM DRIVERS)

- Successional transitions are managed in areas where the absence of disturbance, especially fire, reduces habitat diversity (e.g., Douglas fir overtaking oak woodland, coyote bush encroaching on grasslands) (DISTURBANCE REGIMES)

- Sediment transport to tidal marshes allows them to keep pace with sea level rise (HABITAT-SUSTAINING PROCESSES)

CONNECTIVITY enables movement of materials and organisms

What is it? Connectivity refers to linkages between habitats, processes, and populations across a landscape. This includes both the distribution of resources and habitats in relation to one another and the ability of organisms to move through the landscape (permeability). Connectivity allows for movement of individuals and rearrangement of species assemblages across a variety of scales and connects physical processes such as sediment transport across habitats.

What are the key elements?

LINKED HABITAT PATCHES: Habitat distribution that allows for species movement, exchange of resources and gene flow between habitat patches

OPTIONS FOR SPECIES AND HABITAT RANGE SHIFTS: Space and mechanisms for species and habitats to move to as their ranges shift, including accommodation space

GRADUAL TRANSITIONS: Soft edges between habitat types to support ecotones

HABITAT CONNECTIONS ACROSS GRADIENTS: Expression of habitats across gradients such as salinity and temperature

LANDSCAPE COHERENCE: Habitats that are organized in a way that supports desired processes and ecosystem functions, including the ability of individuals to navigate within the landscape

How does it contribute to resilience?

Connectivity provides the connections, space, and physical and biological gradients needed for species to move in response to changing conditions. This allows organisms to escape unfavorable conditions, take advantage of redistributed or newly available resources, recolonize areas after a disturbance, and exchange genes between populations (LINKED HABITAT PATCHES, HABITAT CONNECTIONS ACROSS GRADIENTS, LANDSCAPE COHERENCE). As a result, habitats can shift, species can adapt, and communities can reorganize as conditions change (OPTIONS FOR RANGE SHIFTS). Gradual transitions between habitats where ecotones are appropriate can provide opportunities for acclimation and adaptation (GRADUAL TRANSITIONS). The appropriate degree of connectivity varies by context (e.g., summer pools in otherwise-dry streams are inherently low-connectivity features). In general, a resilient landscape should be connected but not overly so, in order to maintain the integrity and identity of discrete landscape elements and preclude transmission of stressors (e.g., dispersal of invasive species) between different parts of the landscape (see redundancy, page 22).

EXAMPLES
from Resilient Silicon Valley vision:

- Functionally connected stream riparian patches act as movement corridors for wildlife (LINKED HABITAT PATCHES)

- Open space areas and habitat patches of any size throughout the landscape serve as stepping stones and seed sources for colonization (OPTIONS FOR RANGE SHIFTS)

- Gradual estuarine-terrestrial transition zones allow marsh animals, particularly small mammals, to escape flood waters (GRADUAL TRANSITIONS)

- Habitats occur at varying distances from the Bay and the coast along temperature and moisture gradients (EXPRESSION OF HABITATS ACROSS GRADIENTS)

- Undiverted stream flows maintain natural cues for migration of anadromous fish (LANDSCAPE COHERENCE)

DIVERSITY AND COMPLEXITY provide a range of options

What is it? Diversity (the variety of landscape features) and complexity (including the spatial configurations and interactions between these features) together capture the physical and biological variability at nested scales within the landscape, as well as the interactions between different components. These concepts are considered together because of their interrelated and overlapping nature.

What are the key elements?

RICHNESS OF LANDSCAPE FEATURES: Landscape-scale diversity of habitat types and connections between different habitat types; physical heterogeneity in topography, groundwater, soils

WITHIN-HABITAT DIVERSITY AND COMPLEXITY: Site- or habitat-scale vegetative diversity (e.g., in species, structures, or height) and physical heterogeneity (e.g., in microhabitats, microtopography, and microclimates)

DIVERSITY IN STRATEGY AND APPROACH: Response diversity and a diversity of life history strategies both within and between species

GENETIC AND PHENOTYPIC VARIABILITY: Diversity in genes and traits within species populations

How does it contribute to resilience?

Diversity and complexity help maintain the variability necessary for species adaptation and evolution by supporting a range of responses to a heterogeneous and dynamic environment (GENETIC AND PHENOTYPIC DIVERSITY, DIVERSITY IN STRATEGY/APPROACH). Diverse habitat mosaics can support more niches, bolstering biodiversity and promoting alternative life-history strategies (RICHNESS OF LANDSCAPE FEATURES). Complex landscapes, supporting a variety of microclimates and microhabitats, provide individuals with opportunities for acclimation or refuge during disturbance events and extreme conditions (WITHIN-HABITAT DIVERSITY).

EXAMPLES
from Resilient Silicon Valley vision:

- The landscape supports diverse habitats, including oak woodland, grassland, chaparral and scrub, forest, freshwater wetland, streams, and baylands (RICHNESS OF LANDSCAPE FEATURES)

- Willow groves contain diverse vegetative structure and age classes (WITHIN HABITAT DIVERSITY AND COMPLEXITY)

- Robust populations of different species are present that respond to fires in different ways (e.g., re-sprouters and fire germinating seeds) in fire-prone chaparral and forest systems (DIVERSITY IN APPROACH)

- Sufficiently large populations of key species (e.g., checkerspot butterfly, steelhead) and sufficient habitat diversity (e.g., host plants, stream conditions) support within-species diversity (GENETIC AND PHENOTYPIC DIVERSITY)

REDUNDANCY provides insurance against loss

What is it? Redundancy refers to the presence of multiple similar or overlapping elements or functions within a landscape.

What are the key elements?

- ·····➤ STRUCTURAL REDUNDANCY: Discrete habitat patches and an abundance of key structures within habitats; includes isolation or modularity between habitat elements

- ·····➤ POPULATION REDUNDANCY: Distinct or disconnected populations of a specific species

- ·····➤ FUNCTIONAL REDUNDANCY: Multiple species present that support similar or overlapping ecological functions

How does it contribute to resilience?

Redundancy in structures, populations, or functions provides backups, reducing the likelihood that loss or impairment of one landscape element will lead to the loss of an entire species or function (STRUCTURAL REDUNDANCY, POPULATION REDUNDANCY, FUNCTIONAL REDUNDANCY). Discreteness or isolation between habitat elements reduces the risk of habitat loss from certain stressors such as fire, disease, or invasive species, and can encourage genotypic and phenotypic variability within species (STRUCTURAL REDUNDANCY).

EXAMPLES
from Resilient Silicon Valley vision:

- The landscape supports multiple willow groves, increasing the likelihood that appropriate hydrology and sufficient space can be maintained for at least some of the groves to persist (STRUCTURAL REDUNDANCY)

- Steelhead are supported in multiple streams (POPULATION REDUNDANCY)

- Grasslands and oak woodlands support multiple species of burrowing mammals (FUNCTIONAL REDUNDANCY)

6

SCALE

SCALE provides space and time landscapes need to persist

What is it? Scale is the spatial extent and time frame at which species, processes, and functions operate, from the macro to the micro and the daily to the geologic. Like setting, scale determines how the other principles of landscape resilience can be applied within a particular location.

What are the key elements?

········► LARGE SPACES: Areas of sufficient size to accommodate key physical processes and support sufficiently large and diverse populations

········► LONG TIME SCALES: Broad time horizons over which ecological functions must persist under changing and variable conditions, and at which long-range planning occurs

········► CROSS-SCALE INTERACTIONS: Overlapping ecological functions that occur across multiple spatial and temporal scales; management and planning across multiple spatial and temporal scales

24

How does it contribute to resilience?

Consideration of these resilience principles at large spatial scales and long time scales is a key part of applying them in a meaningful way. Sufficient space and time are needed to sustain key biological and physical processes and support the redundancy, diversity/complexity, and connectivity necessary to foster resilience. Large areas provide room to accommodate landscape-scale processes and large, diverse populations (LARGE SPACES). Similarly, planning at broad time horizons prompts consideration of resilience to infrequent but significant disturbances and allows for consideration of lag times between management actions and ecosystem response (LONG TIME SCALES). The presence of species that perform similar or overlapping functions but operate at different scales can increase resilience by providing a diversity of species responses to perturbations (since species operating at different scales will experience and respond to the same disturbance in different ways), while planning at multiple time horizons ensures that short-term actions do not limit long term possibilities for ecosystems to adapt (CROSS-SCALE INTERACTIONS).

EXAMPLES
from Resilient Silicon Valley vision:

- Willow groves of sufficient size support riparian bird communities (LARGE SPACES)

- Projections for sea level rise and consequent habitat conversion are incorporated into in tidal marsh restoration design (LONG TIME SCALES)

- Short term and fine-scale actions and visions link to long term and large-scale planning and visions (e.g., remnant habitat preserved near areas likely to be available for restoration in the future) (CROSS-SCALE INTERACTIONS)

PEOPLE shape landscapes and provide opportunities

What is it? People are part of ecosystems too, and our actions shape landscapes directly (e.g., land conversion) and indirectly (e.g., climate change). As a result, smart stewardship is an essential component of creating landscape resilience. In particular, humans contribute local knowledge, emotional and financial investment, adaptive management, and opportunities for improving biodiversity within our own habitats, from cities and suburbs to parks and farms.

What are the key elements?

➤ ECOLOGICAL ENGAGEMENT: Knowledge of and investment in local ecology by individuals, communities, and institutions

➤ LANDSCAPE INTEGRATION: Opportunities to support ecological functions across urban, suburban, agricultural and open space lands

➤ ADAPTIVE MANAGEMENT: Stewardship of the land in a coordinated, flexible, and informed manner; learning from monitoring, research, and pilot projects

➤ STRESSOR MANAGEMENT: Management of specific stressors that must be controlled in order to maintain desired ecological functions and biological processes

How does it contribute to resilience?

People's ability to envision, implement, and manage resilient landscapes depends upon a deep understanding of place and the political will to invest in and prioritize conservation and natural resource management (ECOLOGICAL ENGAGEMENT). Since ecosystems are complex, unpredictable, and constantly changing, management that emphasizes flexibility and learning – both about the ecosystem itself and the results of different management approaches – can foster landscapes that more effectively anticipate and respond to uncertainty and surprises (ADAPTIVE MANAGEMENT). Successful adaptive management will sometimes require direct amelioration of particular stressors (STRESSOR MANAGEMENT). In addition, human-dominated landscapes harbor potential to foster ecological resilience within the built environment, both in natural areas such parks and open space as well as though biophilic design and green infrastructure (LANDSCAPE INTEGRATION). In exchange, people can derive co-benefits from incorporating ecological functions into our environments, including ecosystem services such as clean water or flood control and a more profound connection to the natural world and to place.

EXAMPLES
from Resilient Silicon Valley vision:

- Opportunities are available for people to interact with nature through parks and education programs (ECOLOGICAL ENGAGEMENT)

- Rain gardens, retention basins, and other water infrastructure are present and linked to larger regional wildlife support and physical processes (LANDSCAPE INTEGRATION)

- Investment in research, pilot projects and monitoring (ADAPTIVE MANAGEMENT)

- Feral cat colonies and other nuisance species are relocated far from areas of core habitat (STRESSOR MANAGEMENT)

Acknowledgments

The Landscape Resilience Framework was developed as part of Resilient Silicon Valley, a project funded by a charitable contribution from Google's Ecology Program. For more information on Resilient Silicon Valley, please visit resilientsv.sfei.org.

Thanks to our Resilience Advisory Team for their input and review of this document: Mark Anderson (The Nature Conservancy), Eric Higgs (University of Victoria), Richard Hobbs (University of Western Australia), and Katie Suding (University of Colorado). Thanks also to members of our Silicon Valley Regional Advisory Team who also provided valuable comments: Nicole Heller (Pepperwood Foundation), Jeremy Lowe (San Francisco Estuary Institute), and Steve Rottenborn (H.T. Harvey & Associates). We would like to give special thanks to Audrey Davenport, Google's Ecology Program lead, whose leadership and vision have made this effort possible.

Further reading

While by no means comprehensive, the following list provides some of the key references relevant to each of the landscape resilience principles.

INTRODUCTION

Adger WN, Hughes TP, Folke C, et al. 2005. Social-ecological resilience to coastal disasters. *Science* 309(5737):1036-1039.

Berkes F, Colding J, Folke C. 2003. *Navigating social-ecological systems: building resilience for complexity and change*. Cambridge, UK Cambridge University Press.

Biggs R, Schlüter M, Biggs D, et al. 2012. Toward principles for enhancing the resilience of ecosystem services. *Annual Review of Environment and Resources* 37:421-448.

Biggs R, Schlüter M, Schoon ML. 2015. *Principles for building resilience: sustaining ecosystem services in social-ecological systems*. Cambridge, UK: Cambridge University Press.

Brand F, Jax K. 2007. Focusing the meaning(s) of resilience: resilience as a descriptive concept and a boundary object. *Ecology and Society* 12(1):23.

Carpenter S, Walker B, Anderies JM, et al. 2001. From metaphor to measurement: resilience of what to what? *Ecosystems* 4:765-781.

Chelleri L, Olazabal M. 2012. *Multidisciplinary perspectives on urban resilience.* Workshop report, 1st edition.

Curtin CG, Parker JP. 2014. Foundations of resilience thinking. *Conservation Biology* 28(4):912-923.

Holling C. 1973. Resilience and stability of ecological systems. Annual Review of Ecology and Systematics 4:1-23.

Resilience Alliance. 2010. *Assessing resilience in social-ecological systems: Workbook for practitioners. Version 2.0.* Available at http://www.resalliance.org/3871.php

Standish RJ, Hobbs RJ, Mayfield MM, et al. 2014. Resilience in ecology: Abstraction, distraction, or where the action is? *Biological Conservation* 177:43-51.

Zavaleta ES, III FSC. 2010. Resilience frameworks: enhancing the capacity to adapt to change. In *Beyond naturalness: rethinking park and wilderness stewardship in an era of rapid change*, ed. David N. Cole and Laurie Yung, 142-161. Washington D.C.: Island Press.

SETTING

Anderson MG, Clark M, Sheldon AO. 2013. Estimating climate resilience for conservation across geophysical settings. *Conservation Biology* 28(4):959-970.

Anderson MG, Ferree CE. 2010. Conserving the stage: climate change and the geophysical underpinnings of species diversity. *PLoS ONE* 5(7):1-10.

Bengtsson J, Angelstam P, Elmqvist T, et al. 2003. Reserves, resilience, and dynamic landscapes. *Ambio* 32(6):389-396.

Heller NE, Kreitler J, Ackerly DD, et al. 2015. Targeting climate diversity in conservation planning to build resilience to climate change. *Ecosphere* 6(4):1-20.

Higgs E, Falk DA, Guerrini A, et al. 2014. The changing role of history in restoration ecology. *Frontiers in Ecology* 12(9):499-506.

PROCESS

Beechie TJ, Sear DA, Olden JD, et al. 2010. Process-based principles for restoring river ecosystems. *BioScience* 60(3):209-222.

Drever CR, Peterson G, Messier C, et al. 2006. Can forest management based on natural disturbances maintain ecological resilience? *Canadian Journal of Forest Research* 36:2285-2299.

Holling C, Meffe G. 1996. Command and control and the pathology of natural resource management. *Conservation Biology* 10(2):328-337.

Suding KN, Gross KL, Houseman GR. 2004. Alternative states and positive feedbacks in restoration ecology. *Trends in Ecology and Evolution* 19(1):46-53.

CONNECTIVITY

Mumby PJ, Hastings A. 2008. The impact of ecosystem connectivity on coral reef resilience. *Journal of Applied Ecology* 45:854–62.

Standish RJ, Hobbs RJ, Mayfield MM, et al. 2014. Resilience in ecology: Abstraction, distraction, or where the action is? *Biological Conservation* 177:43-51.

Thrush SF, Halliday J, Hewitt JE, Lohrer AM. 2008. The effects of habitat loss, fragmentation, and community homogenization on resilience in estuaries. *Ecological Applications* 18:12–21.

DIVERSITY/COMPLEXITY

Anderson, M. G., & Ferree, C. E. 2010. Conserving the stage: climate change and the geophysical underpinnings of species diversity. *PLoS One*, 5(7), e11554.

Carpenter, S. R., & Brock, W. A. 2004. Spatial complexity, resilience, and policy diversity: fishing on lake-rich landscapes. *Ecology and Society*, 9(1), 8.

Elmqvist T, Folke C, Nyström M, Peterson G, Bengtsson J, et al. 2003. Response diversity, ecosystem change, and resilience. *Frontiers in Ecology and the Environment* 1:488–94.

Norberg J, Cumming GS, eds. 2008. *Complexity Theory for a Sustainable Future.* New York: Columbia Univ. Press.

Standish RJ, Hobbs RJ, Mayfield MM, et al. 2014. Resilience in ecology: Abstraction, distraction, or where the action is? *Biological Conservation* 177:43-51.

REDUNDANCY

Ahern J. 2011. From fail-safe to safe-to-fail: Sustainability and resilience in the new urban world. Landscape and Urban Planning, 100(4), 341-343.

Chapin FS, Kofinas GP, Folke C, eds. 2009. *Principles of Ecosystem Stewardship: Resilience-Based Natural Resource Management in a Changing World.* New York: Springer.

Gunderson LH, Holling CS, eds. 2002. *Panarchy: Understanding Transformations in Human and Natural Systems.* Washington, DC: Island Press.

Nyström M. 2006. Redundancy and response diversity of functional groups: implications for the resilience of coral reefs. *Ambio* 35:30–35.

SCALE

Kerkhoff AJ, Enquist BJ. 2007. The implications of scaling approachis for understanding resilience and reorganization in ecosystems. BioScience 57(6):489-499.

Nash KL, Allen CR, Angeler DG, et al. 2014. Discontinuities, cross-scale patterns, and the organization of ecosystems. *Ecology* 95(3):654-667.

Peterson G, Allen CR, Holling C. 1998. Ecological resilience, biodiversity, and scale. *Ecosystems* 1(1):6-18.

Standish RJ, Hobbs RJ, Mayfield MM, et al. 2014. Resilience in ecology: Abstraction, distraction, or where the action is? *Biological Conservation* 177:43-51.

PEOPLE

Alberti M, Marzluff JM. 2004. Ecological resilience in urban ecosystems: Linking urban patterns to human and ecological functions. *Urban Ecosystems* 7:241-265.

Gunderson LH. 2000. Ecological resilience - in theory and application. *Annual Review of Ecology and Systematics* 31:425-439.

Spirn AW. 2012. *Ecological urbanism: a framework for the design of resilient cities.* Available at http://annewhistonspirn.com/pdf/spirn_ecological_urbanism-2011.pdf.

Tompkins EL, Adger WN. 2004. Does adaptive management of natural resources enhance resilience to climate change? *Ecology and Society* 9(2):10.

www.ingramcontent.com/pod-product-compliance
Lightning Source LLC
Chambersburg PA
CBHW041731210326

41598CB00008B/844